Published by PKI Press
Waltham, Massachusetts USA

Library of Congress Cataloging-in-Publication Data

Best, Don, 1949-
  Moonlight on the Amazon : truth-telling & laughter from a missionary life / by Don Best.
    p. cm.
ISBN-13: 978-1-931248-08-2 (hardcover)
ISBN-10: 1-931248-08-7 (hardcover)
1. Best, Don, 1949- 2. Missionaries–Amazon River Region–Biography. 3. Missions–Amazon River Region. I. Title.
BV2853.B7B47 2009
266.0092--dc22
[B]

2009008636

First edition
Printed in the United States of America

Moonlight on the Amazon

Truth-Telling & Laughter
from a Missionary Life

by Don Best

FYI

*The essays, letters and poems in this book
speak of real-life people and events, except for
The Fox, the Bird & the Monkey, which is folkloric.
In a few cases I've changed people's names in order
to guard their privacy, and keep from being sued.*
Just joking…

Contributors

Photography: Betty Best
Edson Lee
Don Best
Jonathan Bennett
Josiah Huber
Mark Burdette

Photo Editing: Betty Best
Gabriel Caesar
Don Best

Page Layouts: Don Best
Gabriel Caesar

Cover design by Betty Best
Special thanks to Fabio Colombini for the photograph of the macaw on page 14.

For Betty, my beautiful wife and best friend ever.

Moonlight on the Amazon

Preface

The Amazon is a land of rough edges and wild colors, of raucous sounds and pungent smells, of extreme heat and unrelenting rain.

Yet somehow we love it!

That's because it's also a land of warm hearts and passionate faith.

I think the reason the Holy Spirit so dearly loves it here is because the Holy Spirit is here so dearly loved. In the midst of this vast and troubled garden, God truly inhabits the praises of His people, just as He promised.

Despite our many weaknesses as missionaries and our clumsiness as foreigners, God enables us to be of use. For this we thank Him every day! And also at night....

Tom Best

The Land of Thanksgiving

I'm not embarrassed to tell you, old friend, that the very first thing we did when we got to the Amazon, getting off the plane, was to kneel down at the edge of the runway and kiss the ground. It was one of those spontaneous acts of gratitude that just pops out of you sometimes – a release of pure thanksgiving, a prayer without words. And that, so far as I can tell, may be the best prayer of all.

The Lord brought us through more than I can say, more than I can even remember, to deliver us there that night. Miracles of healing. Miracles of provision. Miracles of timing.

So we remained there on our knees for a while, just pouring out the thanks, till it dawned on us finally that we were being watched. I remember looking up and seeing the Brazilian co-pilot and two or three other guys eyeing us from a distance, wondering what stripe of *loucura* this might be.

As you probably know, it takes a lot of weirdness to pop the eyes on a Brazilian, since they're the coolest, most nonchalant, most unassuming people in the world. But hey, we managed to do it right off the plane!

By then, of course, the moment was lost to us. Like Adam and Eve, we suddenly realized we were naked, and should probably go do something carnal, like take care of the baggage.

But thinking back on that moment by the runway, I can imagine no other place I would rather dwell than in the land of thanksgiving. To be wildly and selflessly grateful in every moment, to be giving thanks 24-7, regardless of the circumstances, is to live a life of unceasing prayer and unbroken fellowship. I think this is what the Bible means when it says that David was a man after God's own heart. I think his heart so overflowed with gratitude that it often broke out into songs and poems and dancing without him giving it much thought.

David's wife, full of scorn and sterile seeds, thought it unseemly for the king to dance before the Lord. But God must have loved David's exuberant dance more than all the bulls and rams that were ever roasted on the altar.

To break out in song and dance before the Lord. That's what I want! To worry not what the world might think. Maybe that's what Jesus was trying to teach us when he stood the little child in their midst and said, "Here, be like *this*."

Lose your sophistication.
Lose your self-absorption.
Dwell in the land of thanksgiving,
In the province of joy!

I am not embarrassed to tell you, dear friend, that there have been other incidences here at the mission, just since our arrival, of a rather peculiar nature. For example, the story's told that when the dry season finally broke last year – after eleven weeks of parching, unimaginable heat – a certain missionary was seen out in the courtyard dancing in the rain. No shoes. No

shirt. No umbrella. In his shorts alone, doing a very poor imitation of Gene Kelly.

Some might speculate that he was suffering from culture shock or heatstroke. But I, being very well acquainted with him, can tell you that it was gratitude that led him out into the courtyard and joy that set his feet to dancing. And suddenly, before he knew it, he was in the land of thanksgiving.

Perhaps you will agree with me that dancing in the rain is not such an unreasonable act of worship, especially when you stop to consider this: There is no guarantee here that the trade winds will shift each year and bring back the rains.

No guarantee at all.

And if they fail to come, the Amazon will quickly die, and we along with it.

If that's not enough to move one's heart to praise, I wonder what would?

Snowstorms in the Amazon

I believe that God is faithful to teach us, in both the big things and little, if only we're willing to learn. Just this morning I was roused from my work by a ruckus in the courtyard below my window. And walking over to find its source, I could see below me young Aaron Fuller and Priscilla Huber, and Christina-what's-her-name – and two or three other little kids whose names I haven't learned – wheeling and squealing around the courtyard.

From some place or other they'd gotten some chunks of Styrofoam – the kind of foam blocks that are used to pack computers and stereo gear – and these had become the happy focus of their play. Each kid had a chunk of foam in one hand and a stick in the other, which they scraped across the surface of the foam, breaking off little bits and flakes. Thus, as they paraded around the yard, more or less in file, each left a flurry of white plastic dust flying in his wake, which swirled about in the breeze and littered the grass.

"What the heck are they doing?!" I fumed. "Making a mess like that." Faster than you can say "hissy fit," I left my coffee cup on the windowsill and thundered down the stairs to put a quick end to their game and scold them for their thoughtlessness.

But as I burst out into the courtyard, the kids wheeled about and came right at me, passing close enough now so that I could hear them, gleefully chanting: "It's snowing! We're making it snow!"

And that put the brakes on my sandals, and also my tongue. Imagine that! Snow in the Amazon. Even as the parakeets chattered in the trees above us and the mangos hung ripe for the picking. Snow in the Amazon!

I stood there as in a dream and watched the small white flakes settle on the patio around me, and even on the tips of my sunburned toes. And I thought back to those wonderful mornings in New England, when God's first dusting of snow would settle on the still-green grass.

And I imagined for just an instant that the flakes felt cold across my toes, and that if I watched them for just a moment more, they would melt away, as snowflakes always do.

Because my vision had shifted, and my memory stirred, I was, for a beautiful moment, teachable. *O God*, I thought, *save me from being so utterly adult. Give me yet the eyes of a child, that I might play again, that I might see your wonders…even in a block of Styrofoam.*

And should the question ever be put to me, be it tomorrow, a year from now, or ten: "Does it ever snow in the Amazon?"

"Why sure," I'll say, "If you have the eyes to see…."

"I tell you the truth, unless you change and become like little children, you will never enter the kingdom of heaven." — Matthew 18:3

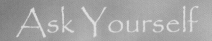

Ask Yourself

What do I own that the Lord has not given me?

What do I aspire to that the Lord cannot accomplish in me?

What evil can approach me that the Lord cannot repel?

What wrong have I ever done that the Lord cannot pardon?

What do I know of love that the Lord has not taught me?

Starlight

"I know that God has much greater people than me working for Him in the Amazon. Pastors and pilots and doctors. But sometimes, out in my village, a beautiful sense of meaning comes over me. On a moonless night – you know, we have no electric lights there – you can see all the stars you want. And sometimes, walking along the edge of the river, I look up and sense that God is looking down at me, and that I am starlight in His eyes."

— Carlos José Miranda
Volunteer River Pastor
Ilha do São Miguel, Brazil

Volunteer Pastor Carlos Miranda, his wife, Isaneide, and daughter, Débora. Their small church, on the river island of São Miguel, has planted seven new churches in surrounding villages and is busy taking the Gospel into others.

Mostly in the Doing

It's sad but true, old friend, that our life stories don't always have happy endings. We wish it were otherwise. And we always pray for the best. But sometimes the tale ends sadly.

I'm not talking here about physical death, which isn't necessarily sad. I'm talking about failures among the living.

I see in the course of my own life, and many others, how disobedience invariably leads to failure. To put it another way, how sin always, unavoidably, exacts its consequence. Yes, there is forgiveness, even for the worst of sins. But the physical and emotional consequences must be met.

I will never forget something that my pastor and good friend Michael Gantt once taught me. He said: "Sin always takes you further than you wanted to go…always keeps you there longer than you wanted to stay…and always costs you more than you wanted to pay."

I am painfully tuned to these thoughts this morning because one of my disciples is grappling with the failure of his marriage. His wife has kicked him out of the house, taken away the children, threatened him with assault charges, and is hell-bent on divorce.

Over the first five months of my relationship with this man, meeting with him at least once a week, he never mentioned to me that he had serious problems with unforgiveness and anger, or that his marriage was on the brink of ruin. Never once. Thus, there was a grievous lack of transparency – or, to put it bluntly, truth-telling – from the very beginning.

When the truth finally did stumble and stutter out into the light, it was very late in the game. Nonetheless, we began to pray and to look into God's Word, and to God's people, for counsel. I believe with all my heart that had this man acted on the counsel he was given in those decisive days, he might yet have saved his marriage.

Alas, he would listen carefully to the counsel of the Word and to the counsel of the saints, agree with us that it was all true and sound and beneficial, and then – because of pride and unforgiveness – refuse to do even the smallest part of it.

I must tell you that I grew so frustrated with him, week after week, that I began to pray for patience. And sometimes found myself without it.

"Why didn't you do what was asked of you?" I would ask him, time after time.

He would shrug and grit his teeth, and offer lamely, *"Não foi possivel."* ("It wasn't possible.")

What he really meant, of course, was, "Because of my pride and hard-heartedness toward my wife, I won't go see her. I know that God says I need to humble myself, and ask forgiveness, but I can't get myself to do it."

"You must kill your pride," I told him. "Else your pride's gonna kill your marriage. You must be obedient to what the Lord says about repentance and humility and forgiveness. When the Word says that you must forgive your wife, and she you, it's not a suggestion. It's a commandment."

But never once did he truly repent of it.

Never once, so far as I know, did he weep over the hash that he'd made of his marriage and family.

Never once did he humble himself and go back to her.

Never once did he do anything to show her that he still loved her.

And so, old friend, the question remains – same as it's always been: Knowing the Word, and the truth, and the law, are we going to obey God or not?

It's commonly taught inside the Christian church that when a principle or commandment is repeated several times in the Scriptures, it's because God – knowing how slow and stiff-necked we are – wants to drive the point home. Being slow and stiff-necked myself, I believe it's true. Thus, I find it very interesting that our Lord Jesus, in His last few hours on earth, hammered home the importance of obedience over and over and over again. It was fixed before Him like nothing else – like the cross itself.

Obedience

Obedience

Obedience…

John describes those last precious hours to us in great

detail, beginning in the thirteenth chapter of John.

"If you love me," Jesus tells us, "You will obey what I command" (John 14:15).

Just a few verses later, He repeats the thought, "Whoever has my commands and obeys them, he is the one who loves me" (John 14:21).

Judas (not Judas Iscariot) interjects a question at that point that would seem to steer the conversation off in a different direction. But Jesus brings us back to the crucial point: "If anyone loves me, he will obey my teaching," He says in verse 23. Then, just two verses later, He flips the idea upside-down. "He who does not love me will not obey my teaching."

I can just imagine Peter leaning over to John at that point and whispering, "Boy, Jesus is really big on this obedience stuff, ain't He?"

Well, yeah!

But Jesus wasn't through yet. After describing the relationship between the vine and the branches, He drives the point home again, "If you obey my commands, you will remain in my love, just as I have obeyed my Father's commands and remain in His love" (John 15:10).

Finally, in verse 14 He repeats it one last time, as personal and as passionate as He can make it: "You are my friends if you do what I command."

Well...what is it in those verses that we don't understand? Did Jesus need to say it a *seventh* time for me to finally get it?

The world is filling up fast with books and CDs that seek to explain and interpret and expand on the Scriptures. And I have no argument with those works. But frankly, my problem is not in understanding the Ten Commandments or the other teachings of my Lord. I think these have been made plain to me.

My problems, like those of my wayward disciple, lie mostly in the doing.　✂

To Clara, Wherever this May Find You...

Dear Clara,

We go back and forth, you and I. Round and round. Fencing, as it were, with well-tooled words.

But death will surely put an end to all debate. *Especially to the pleasant lie that all roads lead to heaven*.

I can hear you now, sputtering, sounding off about how self-righteous and judgmental I am.

But I argue not for Jesus here, or for any specific view of heaven. (Though I might lapse into that at any moment.) Only that Truth exists, and must, by its very nature, be *exclusive*.

I understand from our earlier fencing – which sometimes felt like *boxing* – how upset you get by the notion of exclusivity. If you were any less a lady, you would probably have labeled me "closed minded" or "bigoted."

But I must, with all reason and love, parry your philosophy that heaven is whatever we deem it to be. As though we could somehow shape it through the power of our own will or the desires of our imagination.

I concede you the point that heaven may or may not exist. And if it does exist, that it could be in line with what the Buddhist or the Muslim or the Confucianist holds, rather than with my own Christian beliefs.

But the notion that heaven is whatever I concoct it to be establishes *me* as God and Creator. And this I know, with absolute certainty, I am not.

"Your truth may work fine for you," you tell me. "But I have my *own* truth. And both are equally valid."

Ah, but what you're talking about there is *opinions*. Not truth.

You will perhaps remember from your college days that gold has 79 electrons swirling around its nucleus. From your physics and chemistry classes you'll understand that gold cannot possibly have more than 79 electrons. For if it had – say 82 – then it would be lead, not gold. Nor can gold have any fewer than 79 electrons, for then it would be platinum or osmium or iron.

So you see, I may believe with all my heart that lead is gold. I can paint it yellow, fashion it into fine jewelry, and deem it precious with all my thoughts and words and actions. I can even try to persuade the goldsmith and the banker. And yet, lead will always remain lead. Because it was created so. Immutably.

So why, dear Clara, understanding all this in the realm of physics, does your theology so stubbornly reject the possibility – yea, the likelihood – that God has also established truth in the spirit realm?

I will take the risk here of offending you and suggest that you have embraced tolerance as your God, and thus find everything tolerable, but nothing quite true.

Imagine how popular Jesus would be had He simply watered down the truth about truth. Nothing has engendered so many enemies to Him – then and now – as His unabashed declaration that, "I am the way and the truth and the life. No man comes to the Father except through me."

"I am the gold," He is saying. "And there is no other."

How offensive that is to our flesh, our pride, our political correctness. What we really long to hear is that we are all pretty good Joes and Janes, and that by virtue of a pleasant attitude, an inclination toward good deeds, and perhaps some measure of self-determined spirituality, we will all of us end up in heaven.

What a pleasant lie that is! And hence its popularity.

But Jesus cannot deny himself. "I am the way," He maintains. "And there is no other."

Your distant friend (geographically speaking only),

P.S. With or without our endorsement, the bees still make honey….

Hazardous Straits

O Lord of the hazardous strait,
  Be Thou my trusted pilot.
O Lord of the difficult passage,
  Be Thou my comfort and confidence.
O Lord of the long, cold night,
  Be Thou my warmth and dawn.
O Lord of the perfect storm,
  Be Thou my perfect calm.

"The light shines in the darkness, but the darkness has not understood it." – John 1:5

Karaoke Nights

We lay awake last night and listened to a karaoke party across the street, which wore on and on, till about three in the morning....

You must understand that there's no glass in any of the windows here in the Amazon, and that Brazilians are exceedingly fond of amplifiers. So we had no choice but to listen, and to pray that perhaps they'd blow the amp.

As the minutes wore into hours, the revelers got drunker and louder. One after the other they assaulted the poor microphone, wailing into the night. One of them crooned an unbelievably smaltzy version of "Feelings", mangling the English so badly that we didn't know whether to laugh or weep.

Just when we thought it could get no worse, it suddenly did. One of the girls in the party, egged on by her friends, got hold of the microphone and began to sing her very own and never-to-be-repeated (we pray) rendition of "Material Girl." Despite her jarring lack of rhythm and the fact that she was tone deaf, she bellowed it right out.

For a split second then, I sensed an awful disturbance in *The Force*. Outside in the courtyard, I heard the crickets quit chirping. Dogs up and down the street began to wail in horror – had it been possible, I believe several of them might have committed suicide. Though there was not a breath of wind in the air, the acacia tree outside our window sighed heavily and began to shed its blossoms.

For the first time ever, I silently thanked God for my earwax problem. "She's the worst I ever heard," I told my wife, who had pulled the pillow over her head. Indeed, to me, it sounded like a pig being slaughtered by a very clumsy butcher. Or, at other moments, like sheet metal being ripped through a table saw.

When the girl had finally, mercifully finished, it seemed to me as if the whole of creation breathed a sigh of relief. Then, unexpectedly, we heard her friends begin to applaud her, and with loud, exuberant voices demanded an encore.

"Mais uma vez! Mais!"

My sleepless, sardonic wife poked her head out from under the pillow, and with the most wonderful deadpan imaginable, quipped: "She must be REALLY pretty."

✺

Epilogue: Along with tongue piercing and telemarketing, karaoke may be man's worst invention ever....

What the Brazilians say:

"I have no trouble believing that Jesus turned water into wine. In my own life, in my own home, He has turned whiskey into new furniture!" – *Testimony of a former drunk*

"No one throws rocks at a mango tree that bears no fruit." – *Why Christians come under spiritual attack.*

"I want to lay my head on your chest, Jesus, and hear your heart beat. I want to hide your shoes, so you can't go away...." – *From a Brazilian worship song*

On the River Curuá

The hottest moment in world history occurred just after noon, on Saturday, March 12, 2005, one degree south of the equator, nine days before the equinox, near the little river port of *Curuá*, Brazil.

I have no way to prove this outlandish claim, save the testimony of the crew. No data from thermometers or humidistats that might establish the record with Guinness.

So I'll simply tell you as it happened, old friend, as a tribute to the crew and a marker for my memory....

$\mathcal{I}$ remember looking back across the deck of the *Intimidade* that afternoon and seeing the bodies draped like corpses in their hammocks. All conversation and idle movement had ceased. All appetite was lost. The weather deck above us – built primarily for shade – had become a plate of superheated steel, radiating misery onto the deck below.

In the last hammock aft lay my friend Ronnie Yother, a man of big stature and even bigger faith. He was stripped down to his boxers and a pair of green sunglasses, which were cocked back over his head at rest. His wife, Sandy, hovered on a wooden stool beside him, applying wet cloths to his face and chest, whispering little words of encouragement. Though Ronnie is a veteran of the tropics, and had been dutifully chugging water and water and more water all morning long, the heat had finally laid him wheezing on his back. It occurred to me then – not just in theory, but as a real and present danger – that men can die of heat. I don't mean die of thirst – we had plenty of drinking water in the tanks. I mean die of heat. *Cooked alive.*

Through the gaps between our hammocks I could see the empty dock alongside us and a pair of wooden freighters tied off on the other side. No dock workers. No vendors. No dogs. Even the vultures had fled.

Though I have lived a good portion of my life in the tropics, and experienced the red-hot deserts of Mexico and Southern California, I have never felt the air so deadly still and stifling as it did that moment. Apart from my own tired breath and the languid current in the river, it seemed to me as if all molecular activity had glummed to a stop.

I poured a cup of water across my forehead and felt it run down the sides of my face, back across my naked shoulders and into the sweaty fabric of the hammock. "Oh, Lord, send us a cloud…a breeze…" I whispered aloud. "The memory of something cool."

It was at that very moment – the hottest in human history – that I heard movement on the deck below. Then saw Pastor Reinaldo, our pilot, come lumbering up the ladder. His light-gray T-shirt had turned dark with sweat and sagged down around his waist like a sack. His broad, fire-tested face sparkled with sweat. His tongue wagged out the corner of his mouth a little, panting for air. But there was no complaint in his eyes, nor murmur in his voice.

Come to think of it, I have never heard Reinaldo complain about anything and am convinced that he has learned the deep and beautiful secrets described in Philippians 4. He stood briefly beside my hammock, with one hand resting on its cords. With the other he palmed the sweat off his forehead and shook it off onto the deck. "I want to move the boat," he said, economizing every word. "Somewhere up river. Someplace cooler."

I nodded to him. He nodded back, then turned and went back down the ladder and into the pilothouse.

A moment later I heard the diesel fire and saw Isaque the mate moving along the dock, throwing off the lines.

When the *Intimidade* swung out into the river and we felt that first cool sweep of air moving across the deck, every heart responded in common with little cries and sighs of thanks. "Oh, thank God. *Thank you*."

"Couldn't we just keep on cruising like this forever?" mused Corazon, from the depths of her hammock. "Or if not forever, till dark anyway?"

"If diesel weren't so expensive," I muttered, "we could cruise all night. But they're asking 2.35 a liter…"

"It's going to rain," someone chimed in. "You'll see."

We cruised up the *Rio Curuá* for five or six kilometers, generating our own little breeze. Finally, Reinaldo brought us around a bend in the river and snugged the boat in close to the northern bank, under the broken shade of a Brazil Nut tree.

With universal regret – but no complaining – we heard the engine stop and felt the *Intimidade* rock slowly to a rest. The heat, like a living thing, must have been stalking us along the banks of the river and found us now in our weakness. Prior to that very moment I had never thought of the sun as my enemy and the wind as my friend. But so I perceived them now.

We settled back into our hammocks with a certain resignation, determined to tough it out till sunset. I daresay that everyone on board, down to the geckos in the galley, was earnestly praying for rain.

Despite Reinaldo's earlier warnings about snakes and other unpleasant things that lived in the river, I felt inclined again to throw myself over the railing and take my chances in the water. Instead, I dozed off into a thick, oppressive slumber…full of hot, unpleasant dreams.

It was after three when I finally awoke, and knew immediately, joyously, that the light and air were shifting in our favor. I lay there in a stupor, gazing out across the river, sensing that something beautiful was gathering in the east. Then it spoke to us – the merest suggestion of thunder, but thunder it was – as beautiful to our wilting ears as any poem or music ever scribed.

Anticipation stirred along the deck and quickened our frazzled spirits. I felt like a child in my yearning, overcome with excitement.

A flight of parrots burst out of the jungle on the other side of the river and flew directly over the boat – an omen – a beautiful herald. Behind them came the merest push of air, not quite a breeze, but the promise of a breeze. A fan of wispy gray clouds spread out above us, doing battle with the sun. Go clouds! Go!

Our hearts, as though linked by a common nerve, erupted in spontaneous praise. I leaped out of my hammock and began to dance across the deck, delighting in the folds of wind that coursed around me. O Glory! I could hear the shouts and praises of my friends around. Of the cooks and mates below. I declare, with no exaggeration, that even the geckos were singing!

The shadows deepened in the east and thundered again. Closer now. Closer. A family of iguanas moved along the farther shore. One of them, judging by his enormous size and grayness, was a very ancient fellow. He stopped and cocked his old head to the east. Waiting. Hoping. Just like me.

Now the wind was coming. The blessed wind. I could hear it marching through the jungle toward us, could see it shake the dusty leaves about and push the palms around. Suddenly it broke out onto the face of the dimpling river and danced through our hot little boat like an angel. Like an angel perfumed with rain.

On the heels of that blessed wind came something even better. I will never forget the gentle rhythm of the rain as it came waltzing through the jungle toward us. I leaned out over the railing as far as I dared, so that I might catch its merest whisper and fill my lungs with its delicious fragrance.

A spray of droplets swept across the forward deck, producing little spouts of steam on the hot steel. I could feel the sun retreating in the west, defeated now.

I shall never forget that first sweet curtain of rain that washed across our boat, our souls, that afternoon. I walked forward, out from under the canopy, and let the rain just kiss my head, my face, my naked shoulders. It felt, in some sense, as though I were being baptized anew and raised from the dead.

The moment produced within me an expression of gratitude so pure, so intense, that it found no need for words.

It was only much later, in the quiet of the following morning, that I jotted down these words:

"Lord, I am a desert, watered by your hand.
    Thank you for your loving care, my gentle gardener.
        By you, and you alone, I live, and bloom, and have my being.
            Lord, I am a desert watered by your hand."

The Great Brazilian Coffee Cake Disaster

No one on earth makes cinnamon coffee cakes like Rarú, our Brazilian housekeeper. And if coffee cakes exist in heaven (which surely they must!), I'll wager that no one there makes them better either.

You can imagine, then, how disappointed I was one early morning to discover that I had carelessly left one of Rarú's delights sitting out on the counter overnight, and found it now completely infested with ants. (These were a species of black ant that we sometimes call "wee jets" because of their tiny size and fantastic speed.)

Standing there in my skivvies, grumbling under my breath, I twice picked up the pan and took a step toward the trash can, thinking to toss the whole thing out. And twice reconsidered.

Having lived in the Amazon for a while now, my perspective on what's edible and what's not has slipped a peg or two. "Maybe I can save it," I reasoned, "I mean, they're *just* ants. It's not like they're *cockroaches* or something." (Surely you've heard that ants bathe regularly and brush their teeth after every meal, haven't you?)

Gently, I took the cake pan by its metal edges, raised it an inch off the marble countertop, and let it go. *Clang!*

The sudden shock roused the ants like an earthquake. Dozens of the little boogers came scrambling up out of the depths of the cake. Some of them, I noticed, were holding their little ears as they bailed out over the rim and ran for their lives.

Rarú – woman of faith; unsung heroine; cook without peers...

Twice more I dropped the pan, flushing out more and more of them. But the fourth time I dropped it, no more invaders came forth.

Still, I *knew* there were some die-hards hunkered down inside the cake, bloated on cinnamon and sugar, probably too fat to move.

Probing gently into the dough with a fork, I confirmed my suspicions. Not many, really, especially compared to the size of their original invasion. But more than a couple…

With quiet determination I cut myself a generous piece of cake, set it on a plate, and slipped it into the toaster oven. Without a second thought, I pushed the **TOAST** button and watched through the glass as the heating elements turned bright orange.

If you think me heartless for baking a few ants alive, you should know that ants of every variety are thriving in the Amazon and constantly warring against humans. And very often *winning.* Moreover, the small number of ants that perished inside my coffee cake that morning pales in comparison to the vast number of worker ants, soldiers, queens and politicians they have at their disposal. Why, I doubt they were even *missed....*

Lastly, I should point out that I succeeded in saving Rarú's wonderful coffee cake – rather than being wasteful of food – which must surely be a virtue.

And in the end, I'm here to report that the cake tasted mighty fine indeed, enjoyed with a fresh cup of coffee. Sure, had I looked in closer, I'd have spotted a gritty carcass or two embedded in the dough. But with my taste buds so happy and attentive to their work, I never really looked.

If I am guilty on any charge at all, it would be that I mentioned none of this to my family, who followed me to the breakfast table a short time later. They agreed with me that Raru's coffee cake, so nicely warmed, was *uncommonly* sweet that day.

$\propto$

The Tenth Beatitude: *Blessed are the flexible, for they shall not be bent out of shape.*

*F*olks don't gab much about the weather here in the Amazon, because the weather is – how else can I put it? – *BORING!*

Basically, you can choose between hot *with* rain (January through July) or hot *without* rain (August through December). That's it, folks. No tornadoes. No hail. No hurricanes. No fog. No snowstorms or blizzards to spice the gab. Just hot-wet or hot-dry.

Though truly boring, the Amazon weather does have one very powerful advantage. You can get up every single morning of the year -- without ever consulting the radio, TV or Internet -- and issue your wife a bulletproof forecast: *"Gonna be another hot one today, honey...."*

The Wealthy Ones

Though his left eye is clouded and never quite in sync with the right, Sebastião sees things very clearly. He and his wife, Hilda, live on an island in the middle of the Amazon River, in the little fishing village of *São Miguel*. They have plenty of fruit, plenty of fish, and plenty of slippery, bright-eyed kids. (Seven, I think…) But most of all, they have plenty of faith.

Sebastião and Hilda welcomed us into their home as though we were royalty, shooing out the children so that we could hang our hammocks in their one and only bedroom. Like other houses along the shore, theirs is raised on pilings – rough-hewn poles sunk into sand. Thus, the floor of their house sits about two meters off the ground.

As we climbed the steps in back and entered the kitchen, I noticed that the wooden floor and walls leaked sunlight between the boards and the metal roof was pocked and streaked with rust. While Betty hung the hammocks, I poked my head out the rough-cut window and took a look around. With the rainy season only half-way through, the river was already slipping up around the house, creating little islands of sand and grass. On these the chickens were nervously congregating, accompanied by a pair of dull orange pigs and a dog that hadn't tasted protein in months.

Sebastião says that come Monday he'll be moving the livestock – including his horse, *Canela* – to higher ground. By the end of the month the river will be a meter deep around the house, cutting them off from their neighbors. At that point they'll have only two ways out: by canoe, which can be conveniently tied off at the back door, or by catwalk. The catwalk (suitably named, I think) is a rickety assembly of elevated boards and stakes that winds its way inland, through the papaya trees, past the latrine, to higher ground.

That night, after a fine meal of peacock bass (*tucunaré*), we turned in early. Despite the mosquitoes, I slept fine for a while, until – sometime after midnight – nature called.

I'm here to tell you, old friend, that rousting yourself

out of your hammock in the middle of the night to use that latrine is like riding Splash Mountain in the dark. Armed only with a flashlight, a roll of toilet paper and the urge to go, I discovered that the catwalk was only two boards wide and sagging under my weight. All around me was the pitch-black expanse of the marsh and the river beyond, filled with all sorts of eerie noises. How could I help but wonder, in such a wild place, if maybe a gator or snake wasn't lurking up ahead? I began humming to myself – *not* a good sign – and to imagine strange and embarrassing epitaphs:

GATOR GOT HIM NEAR THE OUTHOUSE

On the whole, I think constipation might be easier....

⌘ ⌘ ⌘

We were guests of honor the following night at the Paz Church. Though it was raining, the little building was jammed with about 50 adults and too many kids to count. Many of them had walked or paddled for miles to be there, and would walk or paddle home again. We were told that about 80 percent of the adults in *São Miguel* are born-again Christians and that most of them are members of the Paz Church.

Since there's no electricity in the village, we met by candlelight. The cross beams above us were festooned with candles of every sort. White candles and blue, some old and some new. Red candles and brown. Stubby yellow candles that sputtered and dripped. It was a dazzling display of flames and shadows.

Sebastião and Hilda showed us to the front of the church, where a bench had been reserved for us.

Being a white-skinned gringo who speaks clumsy Portuguese and is taller than everyone else, it was easy for me to be inconspicuous.

We sang and sang – praises to the Lord – and

danced some too, led by a nimble and happy guitarist. That the boy was able to get such joyous music out of such a lousy old guitar was proof that the Holy Spirit felt right at home.

Sebastião shared a short message from the Word, which was followed by lots and lots of prayer.

Can you guess, perhaps, the focus of their prayers?

Before you answer, I want to remind you that their village has no electricity, no running water, no municipal sewage system and no schools beyond the fourth grade.

Wait! Don't answer yet. Let me just add that the village has no doctor, no dentist, no drugstore, and is six hours by boat from any place that does.

Wait! The list goes on. The people of *São Miguel* have no retirement plans, no health insurance, no cars, no refrigerators, no telephones, no TVs, no iPods, no food pantries, and no idea what they're going to eat tomorrow if the fishing doesn't go well tonight.

You get the idea.

So what do you suppose they were praying for?

The answer: *diesel fuel*. Diesel fuel for their little mission boat – called the *Journey of Faith* – so they could carry the Gospel to a neighboring island, where the people had never heard about Jesus.

By the time the service was over, Betty and I were so deeply humbled that we could scarcely speak. We managed just enough words to thank them for their kindness, to bless them, and to assure them that the tears in our eyes were from joy, not sorrow. Which wasn't altogether true.

Sebastião found me outside, alone with my thoughts. The rain had stopped. A mighty chorus of bullfrogs was singing along the river.

"You liked the service OK?" he asked.

"It was beautiful," I said. "I'll never forget it. *Ever.*"

Above us the clouds were scudding off to the west and a million stars were out.

"What are the people like where you come from?" Sebastião asked.

"The United States, you mean?"

"Yes."

"Well, before we came here, we lived in New England," I explained.

"And are the people of New England people of great faith?" he wondered.

I felt as if my heart were being squeezed by a pair of

invisible hands. "No."

He stepped in a little closer, the way Brazilians do, and touched me on my arm. "No faith in God?"

The pressure on my heart was pushing tears out the corners of my eyes. "Not really," I admitted. "Not many."

His good right eye strayed into the heavens above, looking for answers. By and by his left eye followed it up. "If not in God, where do they put their faith?"

I shook my head. "I don't know. In themselves I guess. In their stuff."

He cocked his head in a peculiar way. "Stuff?"

I shrugged. "Money. Cars. You know… *Stuff.*"

A long moment of silence passed, with only the bullfrogs sounding. Then he said, "But it all goes back to dust."

"Yes."

Revelation bloomed across his face, then irony. "So *we're* the wealthy ones!" he laughed.

"Yes. *You're* the wealthy ones," I agreed. "Richer than Bill Gates and Warren Buffet combined."

He wasn't familiar with the names, but seemed to catch my drift. "If you should ever go back to New England," he said, "tell them we're praying for them. That they might become people of faith."

Like I said, here's a man who sees things very clearly.   ⤳

Close Encounters of the Amphibian Kind

During a visit to *Alter do Chão*, on the Tapajós River, Tanya woke up in her hammock one night, unsettled by an alien presence. Running her hand across her face, she discovered there was a little frog sitting on her lips. With a "Yipe!" and a quick swipe, she sent the poor fellow flying into space. Apparently she reacted so fast that the frog never had time to kiss her. We know this because – well, it's *obvious*, isn't it? – he remained a frog.

Either that or Tanya wasn't truly a princess….

The Drummer Boy of Renascer

*O*ut on the edge of the world, under the banana trees, the boy had built himself a set of ramshackle drums. From old paint cans and posts he'd fashioned the snare drums and bass. He'd turned the bottom of a rusty old barrel into a cymbal, and a metal dinner plate into a seat.

His name is Diemenson Mendes. He lives in the tiny village of *Renascer* ("Reborn"), where the mission is helping to build a new church.

Before school and after, whenever he gets the chance, Diemenson sits down at his drums and worships the Lord Jesus.

Following a prayer meeting at the boy's house, his mother lured us into the backyard so that we might meet the "church drummer" and see his "*electric* drum set." Filled with curiosity, we splashed and slipped our way out across the yard, which had been drenched by the morning rain.

Neither words nor pictures can quite describe the conglomeration of paint cans, plastic buckets, plates and posts that we discovered underneath the banana trees. Ragged as the materials were, we could see that they'd been assembled with enormous ingenuity and care.

Shyly, Diemenson sat down on the metal dinner plate, took up his rough wooden sticks, and slipped his right foot into a contraption that – *could it be?* – *yes*, the pedal for the bass drum!

Suddenly the boy's sticks were flying across the face of the drums, pounding out a fine, steady rhythm. Then, in a strong, clear voice he began to sing:

(34)

O tempo de cantar chegou!	*(The time to sing has come!)*
O tempo de dançar chegou!	*(The time for dancing's here!)*
E Ele vem, e Ele vem!	*(And here He comes, here He comes!)*
Saltando pelos montes.	*(Leaping through the hills.)*
E Seus cabelos, Seus cabelos	*(And see His hair, His hair)*
São brancos como a neve.	*(Shining white as snow.)*
E nos Seus olhos, nos Seus olhos, há fogo!	*(And in His eyes, His eyes, there's fire!)*
Incendeia Senhor a Tua noiva!	*(Lord, impassion your bride!)*
Incendeia Senhor a Tua igreja!	*(Lord, ignite your church!)*
Incendeia Senhor a Tua casa!	*(Lord, set your house on fire!)*
Vem me incendiar!	*(Lord, come light a fire in me!)*

By the time the song ended, we were so astounded by the Holy Spirit and humbled by the boy's simple heart of praise that we all broke out in cheers and applause. Diemenson, overwhelmed by all the attention, dropped his head and softly began to cry.

We know that God is looking for true worshipers who will worship Him in spirit and truth. There, on the edge of the world, under the banana trees, we had the great pleasure of meeting one. ∝

So You Don't Believe in Miracles, Huh?

Abaniza the cook was brushing her teeth on the back of the mission boat one afternoon when she accidentally spit her false teeth into the river. One of the crew boys valiantly dove in after them and against all odds came up out of the murky water with her teeth in hand!

Now *that*, my friend, is a miracle!

Pirarucú for dinner

Toasting wild manioc

The Fox, the Bird
&
the Monkey

Once upon a time, in the dark green hollows of the rain forest, there lived a mother bird and her five little chicks. Their nest sat high in the boughs of a great acacia tree, as old as the wind....

Well, everything was just splendid and chirping right along, till one morning a fox showed up at the foot of the tree. And he shouted up at the mother bird, "It's breakfast time. And I'm mighty hungry. I think I'll climb up there and eat you for breakfast!"

And the fox began to scratch and claw at the foot of that old acacia tree, making the most awful racket.

Scraak! Scrittle! Screech!

When the mother bird heard this, she was all aflutter and afraid, and shrieked, "Don't come up! Don't come up!"

The fox shouted back, "Well… maybe I won't come up if you throw me down one of your little ones."

The mother bird hemmed and hawed and fretted, and though it broke her heart in two, she threw one of her chicks out of the nest.

And the fox gobbled it up for breakfast and went away… *Smug.*

Well, it should come as no surprise to you that when lunchtime rolled around that old fox was back at the foot of the tree. "It's lunchtime," he shouted up, "and I'm mighty hungry. I think I'll climb up there and eat you for lunch."

And he scratched and clawed at the tree and made the most awful racket. *Scraak! Scrittle! Screech!*

When mother bird heard this, she was all aflutter and afraid, and shrieked, "Don't come up! Don't come up!"

The fox shouted back, "Well… maybe I won't come up if you throw another one of your little ones down."

Mother bird hemmed and hawed and fretted. And finally picked up another one of her chicks, and – though it broke her heart twice over – tossed it out of the nest.

And the fox gobbled up the second chick and went away… *Smug.*

Well, though it breaks *my* heart to tell you this, the same thing happened all over again at dinnertime. The mother bird was so filled with fear that she sacrificed yet another chick to the fox, so she was left now with only two.

And that night there was naught but tears and terror in her nest. And so great was the sadness there that the acacia tree began to shed its yellow petals, like tears, floating to the ground below.

When morning came to that dark green hollow of the rain forest, the fox showed up for breakfast, and shouted up into the tree, "Throw me another chick, or I'm coming up there to get you!"

And the mother bird was all aflutter and afraid, and was on the very very verge of throwing yet another chick down to the fox, when she heard a shout in the distance: "Don't do it!"

And looking round, she saw a little monkey swinging through the trees toward her.

Now, this particular monkey, whose name was "Missionary Monkey," was familiar with the Truth and fond thereof. His mission, inspired by love, was to be a source of hope and courage to all the good creatures in the forest.

"Don't do it!" Monkey shouted, swinging into the acacia tree. "Don't let him take another one!"

"And why not!?" shrieked the mother bird. "He's coming up! The fox is coming up to get us!"

Monkey climbed right up to the edge of mother bird's nest. And leaning in close, filled with the Spirit of Truth, he whispered into her ear, "'Cause foxes don't know how to climb!"

∝

The End

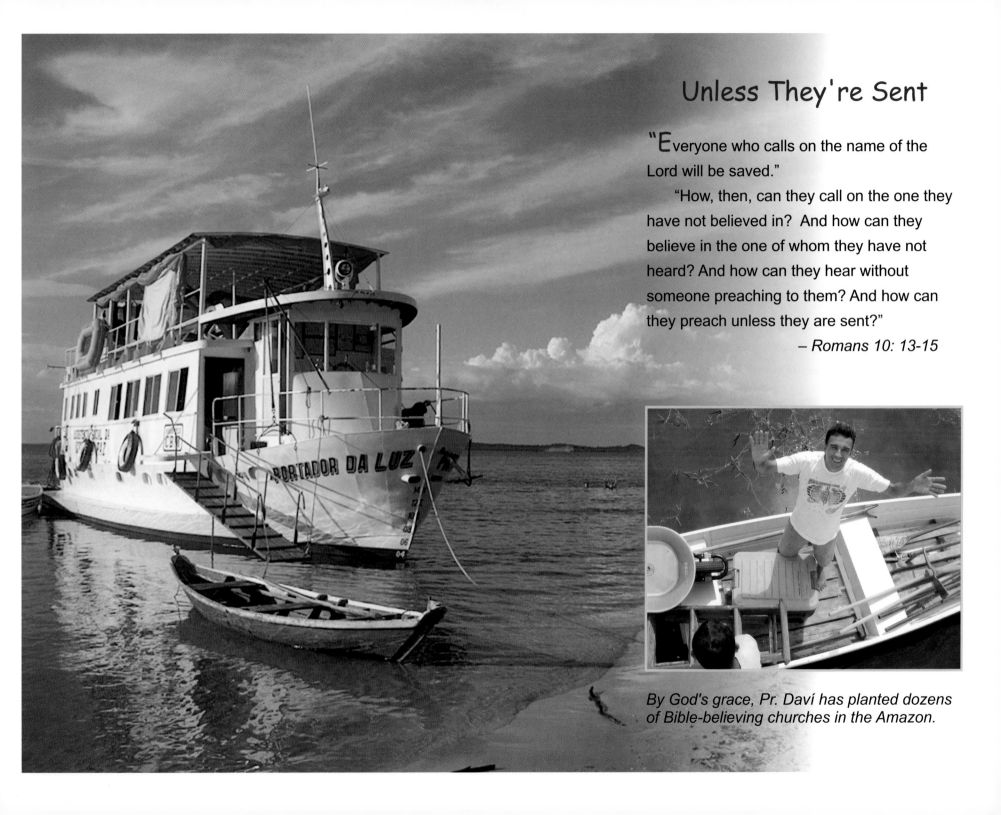

Unless They're Sent

"Everyone who calls on the name of the Lord will be saved."

"How, then, can they call on the one they have not believed in? And how can they believe in the one of whom they have not heard? And how can they hear without someone preaching to them? And how can they preach unless they are sent?"

– Romans 10: 13-15

By God's grace, Pr. Daví has planted dozens of Bible-believing churches in the Amazon.

Faithfully Yours...

Dear Clara,

I was so excited to receive your letter! For a while there, I thought perhaps you'd given up on me and stopped writing. But I see from the postmark that it took your letter *22 days* to get here. I don't know what becomes of the Brazilian mail sometimes – it must have come up the Amazon by turtle.

I found all of your thoughts interesting, Clara, and some of them very familiar.

To me, the best of our gifts is *not* intelligence.

I know, with all certainty, that intelligence will fail me in the end.

Wisdom is better, and certainly worth pursuing. But it too will fail me in the end.

When I reflect on the value of intelligence and wisdom in my life, the yellow flag immediately goes up. Not only will they fail me in the end, but they invite into me the things that God hates most: pride, and its evil twin, self-sufficiency. (Or better said, the "illusion" of self-sufficiency.)

To me, there are three things that are better than intelligence, and better even than wisdom. They are *faith, hope,* and *love*.

The gift of faith, which is given by the grace of God, cannot be won through intelligence, wisdom or good works. It is the only gift that will see me through death, and through judgment.

The Bible says that if you confess with your mouth that "Jesus is Lord" and believe in your heart that God raised Him from the dead, you will be saved.

Try as I may, I find nothing in there about IQs....

Faithfully yours,

"Dear God, help us to remember that many things won't matter in the end...."

Obrigado pelas Rosas

(*Thank You for the Roses*)

I've seen 22 buzzards perched in a single tree
A table built with every leg a different length
A tarantula the size of a hubcap
And a lily pad the size of a tire.
   But I never thought I'd see roses in the Amazon.

I've seen a bolt of lightning stand on the river for three full seconds
An iguana fall out of an acacia tree (and almost hit me)
An old man with no teeth but lots of faith
The world's best coffee…and worst peanut butter.
   But I never thought I'd see roses in the Amazon.

I've seen thunderheads climb six miles high
A mango tree full of chattering parakeets, blue and green and yellow
Seventeen different kinds of bananas,
And pineapples at 25 cents apiece.
   But I never thought I'd see roses in the Amazon.

I've seen a man carry two enormous bags of ice, one on each shoulder, barefooted,
hobbling over hot asphalt…then down the side of the seawall on rickety steps,
over a narrow gangway and onto a river boat,
where he heaved both bags *simultaneously* into an open hold.
   But I never thought I'd see roses in the Amazon.

(44)

I've seen ants gather at my feet, plotting to carry me away
A rat swimming in my toilet
A cricket on my toothbrush
And a white owl sitting on my windowsill.
  But I never thought I'd see roses in the Amazon.

I've seen kids playing happily with nothing but dirt and pop bottle caps
A dead woman lying much closer to me than I wanted, for much longer than I wanted
A parade of religious people three miles long, following a wooden idol through the streets
A man shimmer with the Holy Spirit as he preached.
  But I never thought I'd see roses in the Amazon…

Till this afternoon, that is, when your very red roses arrived!
What a wonderful extravagance! And thank you very much!
We haven't a clue, you know, from where they came.
"Some place down south," our houseboy speculates. "*Way* south."
By plane, of course, for nothing so lovely could have ever survived the road.

You will have guessed, probably, that roses can't last very long in this wilting heat,
Though we pampered them well enough.
But oh what a wonderful sight! And smell! Every minute that they lived.
  Who would have ever thought we'd see roses in the Amazon!

Songs of the Heart

I remember a sad afternoon in June, sitting on the balcony at the mission house in Castanhal. Though the view from the balcony was splendid and the air scrubbed fresh with rain, I was inclined to be miserable.

The big toe on my left foot was deeply infected, oozing pus into a pan of warm saltwater. My intestines churned and burned, and sent me hobbling to the bathroom every 20 minutes. It was worms probably, or maybe giardia. (When they talk about the Amazon in the travel brochures, they never mention the *intestinal* adventures.)

Worst of all was the spirit of self-pity that had cozied up beside me and found a willing ear. *"Poor we…"* it whispered. *"Let us coddle ourselves and commiserate. Only we know how we really feel…right? By and by, if you'll let me linger, I'll introduce you to my good friends, discouragement and depression."*

It was in the very midst of this four o'clock funk that I suddenly heard someone down in the courtyard begin to whistle. I couldn't see the man, but presumed him to be the gardener, for I could hear the sound of his shovel biting into the soil. There was real power in the way he whistled. Not in the tune itself, though that was pleasing enough. But in the easy and unpretentious joy of its expression – the overflow of a happy heart.

He whistled on for three or four minutes and then abruptly stopped. Scarce had his melody left my ears when the housekeeper, working in the kitchen behind me, broke into song.

"Eu tenho um amigo que me ama…" she sang. ("I have a friend who loves me…") *"É Jesus…"*

There was real power in the way she sang – power enough to lift my spirits too – for her voice was bursting with joy. I felt sad when she moved her chores into the bedrooms down the hall, taking her song along with her.

"Ai de mim…" I whined. Poor me.

But there was only a short intermission before yet

another song came booming up the stairs. It was that familiar old chorus, known the world over: "Hallelujah…Hallelujah…Hallelujah…Hallelu…"

I guessed by the voice that it was Pastor Filipe, somewhere on the floor below. There was real power in the way he sang – power enough to lift my spirit too. Not because he was a world-class tenor, but because he so glorified the Lord with his song. It was as if a good sweet well lay within him that could no longer be contained, and must, of needs, now flow and overflow.

After a while, it dawned on me that Filipe had quit singing and that I, quite unconsciously, had started humming the song myself. And noticed further that the spirit of self-pity had fled the balcony, along with its dismal friends.

It was good then, dear friend, to remind myself of this: "The joy of the Lord *is* my strength."

And to ponder a peculiar little question, which I'll share with you here:

Do you sing because you're happy…or are you happy because you sing?

❧

Bats Ridiculous!

On one of our inland mission trips, Patrick went spelunking with two Brazilian pals and found himself standing inside a large cavern filled with hundreds of swirling bats.

His friends, fearful of being hit by one of the fast-flying creatures, hunkered down on the floor.

But Patrick stood his ground, pooh-poohed his friends for their timidity, and began to expound on the multitudinous virtues of bats and their flawless ability to navigate, even in the darkest dark. On and on he went about bat sonar and bat wing design and bat maneuverability, never flinching. To hear him tell it, every last one of them had gone to flight school and graduated with honors.

Just as he reached the climax of his exhortations and praise, a bat smacked him rudely on the cheek, clung there for a startled second, and flapped clumsily to the ground.

Patrick waited until his friends' laughter receded, which took several lifetimes, then calmly observed: "There *is*, of course, always the *class idiot*…."

Accidental Dawn

See the accidental way in which the sun comes up,
  poured out like a palette upon the waiting clouds,
    all gold and plum and scarlet?

See the accidental way its light streams through the mango trees,
  pausing there just long enough to kiss the ripening fruit?

See the accidental way it's found the open window now,
  a pulse, a wave, a dream,
    a gracious beam that skips across the bed to kiss the baby's cheek?

See the accidental baby, born in the midst of a thunderstorm,
  on the scariest day of the year?

That night, when the storm clouds finally cleared,
  an accidental moon rose full above the river,
    and against all odds the baby lay there healthy.

See the accidental father, sitting by his window,
  wondering who to thank…

"For since the creation of the world God's invisible qualities – His eternal power and divine nature – have been clearly seen, being understood from what has been made, so that men are

Antônio's Regret

There is nothing sadder than an old man filled with regret. He is like a story poorly written, that can never be redone. Or wine gone sour in the cask.

No one at the mission will ever forget the bittersweet testimony of Senhor Antônio, who accepted the Lord at 63. Not only was his body in failing state, but so too his family. A daughter lost in promiscuity and rebellion. His son a creature of the gangs, scarred by knives and multiple arrests.

I remember the night when Antônio first came to our cell group – a wee little man of very few words. When the group leader asked him to share his testimony, it was an awful struggle for him, and for a very long time the words wouldn't come. Then, like pulling teeth, he stood up and said, "I was asleep. Now I'm awake. Thank God!"

Just eight words.

Then the tears began to flow. First with him, then, one by one, with the rest of us. Mingled with the joy of his salvation was the bitter knowledge that he'd lived

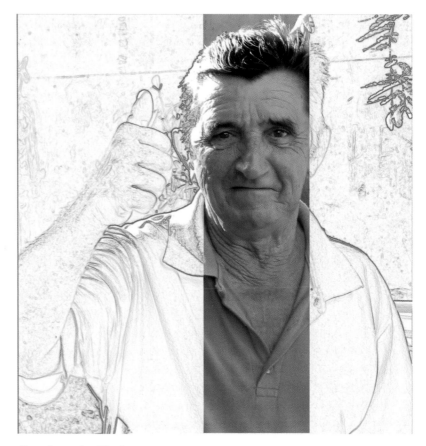

Senhor Antônio

most of his life in darkness, and planted his children there too.

Over the weeks that followed he seemed often to be in tears, or on the verge. I wondered to myself how many of his thoughts and prayers must have begun with the words, "O Lord, I wish that I hadn't…" Or, maybe worse,

"O Lord, I wish that I had…"

Verses like Deuteronomy 11:19 and Proverbs 19:18, commanding us to teach and discipline our children, must have fallen on him like a hammer. How could he follow such commands when his children were grown up now and scarcely at home, when his daughter was in the wild sway of her boyfriends and his son in liege to the gang? "Too late," he must have thought. "Way too late…"

Antônio understood that everything was forgiven him by the blood of Jesus. That the sacrifice was perfect. That there is no condemnation in Christ. But so deep and painful was his regret – especially over his children, and the evil that involved them – it seemed to murder all his joy.

Many of us at the mission joined him in praying for his children. And heard him, more than once, confess his failings as a father: the prayers he'd never offered them; the discipline never applied; the Word never taught.

Together, we are believing that God is going to do a powerful work of restoration in Antônio's family, saving his children. Even as I write this, we have word that his daughter, Suelí, has stepped back a little from her dangerous lifestyle and is showing some interest in the Gospel.

However that may be, we can learn something important from Antônio's regret. Unless God should do an outstanding miracle in time and space, bending the very laws of physics, Antônio will never again have the chance to hold his little girl in his lap, to speak of God's love and faithfulness to her, while her heart is still tender. Those hours, apart from a miracle, have been lost to him forever.

In light of that sadness, let us seize each hour for what it is, a precious and irredeemable gift. Let us recognize the opportunity in every minute, and squander nothing. Thus, should we live to be old, or even not so old, regret will have no rooms to haunt.

⊂×

"Teach my words to your children, talking about them when you sit at home and when you walk along the road, when you lie down and when you get up." — Deuteronomy 11:19

Toiling in Vain

So much talk and so little communication
so much information and so little wisdom
so much sound and so little hearing
so much seeking and so little found
so much money and so little wealth
so much religion and so little faith
so much hate and so little forgiveness
so much doing and so little done

Be honest now! What did you do yesterday of eternal value?

Did you remember, as you hustled through the day,
that God loves you…

So
  So
    Much…

…that He gave you Jesus?

Imagine what we will of our plans and projects;
Apart from Him we toil in vain.

"I am the vine; you are the branches. If a man remains in me and I in him, he will bear much fruit;
apart from me you can do nothing." — John 15:5

Early morning chores in the village of Uruarí...

A Passion for Souls

*J*esus performed many wonderful miracles while He lived among us. But it wasn't primarily to do miracles that He came.

Jesus was an extraordinary teacher – the greatest the world has ever known. But it wasn't primarily to teach that He came.

Jesus came because He has a passion for souls. Because God looked down on a fallen world and saw that men were lost in darkness, enslaved to every kind of depravity. And so sent Jesus, His beloved Son, to set us free.

It was His passion for souls that led Him to the cross.

Love nailed to a tree.

For *me*.

II

Sailing up the *Rio Arapiuns*, Pastor Reinaldo pulled the health boat into *Moaçá*, a village that had never been reached for Christ.

On board that morning was a missionary nurse named Becky Hrubik and three of her children – Beth, Paul, and Deborah. Though the kids were young – Beth, at 12, was the oldest – Becky had wanted them to see first-hand what it was like to work out on the rivers.

They were met on the beach by Flauta, the village leader, who greeted them and led Reinaldo and Becky up the long rough steps to the village. *Moaçá* is a community of about 800 people, spread along the bluffs above the shore. The villagers are mostly fishermen and manioc farmers. The richest among them own some chickens and maybe a cow.

Flauta accompanied Pastor Reinaldo and Becky into the village center, where there was a simple pole building that served as a community center. Beside it sat an aging generator, with strands of weather-beaten wire looping off into the village.

Reinaldo and Becky outlined their plans: to offer a free health clinic that afternoon on board the *Portador da Luz*, then hold an open meeting that night to show the *Jesus* film and share the Gospel.

Flauta agreed, and said they could use the community

Pastor Reinaldo at the helm

center and generator to show their film.

But Flauta's brother, who was the village catechist, was of a very different mind. He had been trailing them like a shadow that morning, full of dark looks and dire mumblings. "Why don't you go away," he snarled at one point. "We don't need any more *religion*."

But Flauta held sway, and the plan went forward.

The clinic was a fine success. Word got out fast – not just in *Moaçá*, but into surrounding villages too – that there was a dentist and nurse attending, with *free* medicine. Villagers lined up early and kept on coming through the long afternoon. For some, it was their first chance ever to see a dentist or talk with a real nurse.

But after sunset, with the clinic over and the people sifting back to their homes, the mood began to change. Something was lurking in the shadows that liked them not.

A little before seven, Pastor Reinaldo and two of the crew members were up in the community center, setting up the projector, preparing things for the evening service.

Suddenly, out of the darkness came a horse and rider. It was the catechist, full of spit and fury. He reined up his horse with one hand and brandished a wooden club in the other.

"Get out!" he shouted at them. "We don't want you here! You understand!? Take your devil boat, and your blasphemies, and get out!"

It was clear from the way he pitched and swayed on his horse that he was extremely drunk and capable of doing serious harm. Out of the shadows behind him, like phantoms, came some of his vigilante buddies, also drunk and bent on violence.

The catechist spurred his horse forward and galloped wildly around the building, slapping his club against the poles. His buddies stumbled in through the doorway, knocking over chairs – intent, it seemed, on wrecking the projector.

Reinaldo snatched the machine off the table and the three of them scurried out the far side of the building. Down the steps to the beach they ran, their hearts pounding, looking back over their shoulders to see if they were being chased.

Safe on the beach – or so it seemed – they stopped to catch their breath.

Becky came running to see what was the matter. When she learned what had happened up in the village, she just thanked God that everyone had made it down safely.

Now a lone figure came hurrying down the steps and across the beach toward them. It was Flauta. "I'm so sorry this has happened!" he blurted. "You must forgive us! You must forgive my brother."

Neither Reinaldo nor Becky could think of anything to say.

"I want you to come back up to the village," Flauta pleaded. "You can set the projector up at *my* house, in *my* yard. I promise you, you'll be safe there."

Reinaldo and Becky looked at one another warily.

"We should pray about it," Reinaldo urged.

"Definitely," said Becky.

So they stood on the beach and prayed aloud, asking God to calm their hearts and give them discernment.

Flauta stood back and eyed them curiously. What little he knew of prayer would have been confined to three or four masses a year, spoken by rote, shrouded in ceremony. To see folks praying on the beach like this, as though Jesus were a living friend, struck him mighty strange and interesting.

After a few minutes, Reinaldo broke off the prayer and looked over at Becky. "I think we should go back up," he said firmly.

Though he was as gentle a man as you could ever hope to meet, violence had played no small part in Reinaldo's life. Working as a river pastor and evangelist, he had been shot at more than once, including one wicked night when a rifle bullet parted his hair. Over his 20-year career on the river, he had been threatened by various catechists and witch doctors – jealous of the Gospel – who had poured sacks of urine and fish guts through the window of his boat…severed the power cable that lighted an evening service…vandalized the building materials that were going into a new church…and dumped buckets of human excrement into a jungle pool that he was using for baptisms. Like the Apostle Paul, Reinaldo counted it all as gain, dwelling not on the opposition or violence that he'd faced, but on the hundreds of Brazilians that he'd led to Christ. *Including many of his worst tormentors.*

Becky took a deep breath and smiled bravely. "Yeah, let's

PAZ missionary Becky Hrubik

go back up. But first, I want to talk to my kids."

She went quickly to the water's edge and climbed aboard the *Portador*, calling her kids forward. She lined them up on the deck, military style – Beth, then Paul and finally little Debbie – giving them each a hug.

 "I want you to know that Mommy loves you very much," she began.

They bobbed their heads up and down, almost in unison. Of their mother's love they had no doubt – *so why this sudden need to say it?*

She leaned in a bit closer to them. It was gravity she wanted to convey, not fear. "No matter what happens tonight, I want you kids to stay here on the boat. Do you understand?"

Again they nodded. *Something had happened in the village, they knew. Something that had not gone according to plan.*

"OK then," she finished. But as she turned to go, her heart drew her back and compelled her to speak again. Looking into their beautiful faces, she said, "If Mommy doesn't come back, *I want you kids to live for Jesus....*"

III

"I eagerly expect and hope that I will in no way be ashamed, but will have sufficient courage so that now as always Christ will be exalted in my body, whether by life or by death. For to me, to live is Christ and to die is gain.*"

– From Paul's letter to the Philippians

IV

As Becky climbed up the steps to *Moaçá* that night, she felt as if she were the luckiest woman in all the world…placed exactly where God wanted her to be…doing exactly what she should.

With perseverance and prayer they showed the *Jesus* film to a good crowd of villagers that night. And afterwards had the enormous privilege of leading Flauta, his wife and two of their daughters to the Lord!

Not long after that, PAZ planted a Bible-believing church in *Moaçá*, which is growing under the favor and protection of the Lord.

V

That's how I want to live my life: *on the steps going back up to the village....*

VI

After hearing the wonderful things that had happened in *Moaçá*, I asked myself, "Where does that kind of perseverance and courage come from?"

And the Spirit answered me: "Jesus has a passion for souls."

"And how does one get that kind of passion?" I prayed.

Jesus replied, as He had to the Pharisees in Mathew 22, "Love the Lord your God with all your heart and with all your soul and with all your mind. This is the first and the greatest commandment."

"And if I should come to love the Lord that way, with all my heart and all my soul and all my mind, what then?" I asked.

For a long time the Spirit did not answer me. Then: "He will teach you two things: *true humility* and a *passion for souls.*"

Suddenly I could see my Teacher there, with the towel wrapped around His waist and the basin in hand. My servant King. Willing to wash my feet, that I might learn how to wash yours. *True humility.*

Looking further, I could see my Savior standing on a hillside above Jerusalem, gazing down upon His beloved city. The crucifixion is only days away. There are tears in His eyes – not for *himself*, mind you, but for *me* – flowing down upon my life. "O Jerusalem, Jerusalem," He laments. "How often have I longed to gather your children together, as a hen gathers her chicks under her wings." *A passion for souls.*

Some things, by the grace of God, are now made clear. The Holy Spirit is bound and determined to conform me to Jesus, be it through conviction, through counsel, through communion, through encouragement, or through healing. If I will love God the way that He has commanded me to love Him, that is, with *all* my heart and *all* my soul and *all* my mind, He will impart to me the very character of His Son: *true humility* and a *passion for souls*.

I think back to *Moaçá*. Of Pastor Reinaldo and Becky climbing back up the steps in the dark, into the very teeth of evil.

Though I would never claim to own such courage, I know now from whence it comes: *From God's own heart, poured into the life of an ordinary believer.*   ⤛

Bugs in the Beans

They say the longer you spend on the foreign mission field, the less picky you get about the food. To wit:

First year on the mission field: "*Gross!* There's a *bug* in my beans. No way I'm gonna eat this!"

Second year: "Hey, there's a bug in my beans. Well, I'll just pick him out and go on eating."

Third year: "Oops, there's a bug in my beans. No problem. I'll just pretend he's not there."

Fourth year: "*Hey, where's my bug?!*"

Without Him...

The boat don't float
The goat don't bleat
The beat don't rhyme
The time don't tick

The eye don't see
The tree don't grow
The foe don't flee
The sea don't roll

The voice don't sing
The king don't reign
The rain don't fall
The ball don't bounce

The ounce don't weigh
The day don't break
The steak don't grill
The pill don't help

The hunch don't pay
The hay don't pitch
The stitch don't hold
The cold don't end

The wound don't heal
The meal don't serve
The nerve don't hold
The bold don't pray

The house don't stand
The band don't laugh
The calf don't rope
The boat don't float
   Nope...

  The boat won't float.

Where Lightning Lingers

I dream of going back on the river, where the monster fish play, where God stands His bolts of lightning on the water like trees of white-hot fire. I've seen them last for three full seconds – an eternity for lightning – blinding every sense, electrocuting the fish.

In my dream, the dolphins always meet us at that special place, where the boat turns south off the Amazon and into the inlet to the great lake. The channel narrows. The water clears. Fans of marsh grass and water lilies reach out from either bank. Herons, first by the dozen, then by the hundreds, rise up from the grass. They sail across our bow in snow-white flights of six or eight, bound for who knows where. Of all God's fine and lovely inventions there is none like rain. But close behind it come the herons, rising into flight.

An hour or two inside the inlet sits a little river village named Alligator (*Jacaré*). Further on is a sister village called Little Alligator (*Jacarézinho*). They tell me – though I've never visited the place – that there is yet another village, named Castrated Alligator (*Jacaré-Cangá*). I'm sorely curious about that name and its origins. Can't help but wonder if the procedure was done *live*. And if so, by *whom*?

Chugging along the inlet, the boat's old Yamaha engine spits and poots, scaring up birds of every sort. Cuckoos and kingfishers. Honeycreepers and orioles. And other birds I've never seen before, whose names I've yet to learn.

Imagine what God could show you of His handiwork if you could put on a pair of rubber waders – the kind that trout fishermen use – and wade up into the marsh grass toward the shore. You would find it alive with every imaginable thing, from fish that spit to dragonflies with 10-inch wings.

The problem with this plan is that you would quickly wade into the realm of the alligators, who lie in the cool mud along the shore. To appreciate them like that, up close, with their beautiful blue armor and awesome jaws, would also give them the chance to appreciate *you*. ∝

Someplace Farther Out

"*I need to move someplace farther out,*" *Zé mused. His eyes sailed out over the river, into the endless expanse of rainforest on the other side....*

"There *is* no place farther out," I told him. "This is *it*…the edge of the world."

He smiled at that, but I could see that his eyes were yearning west, toward some misty little spot undiscovered.

Zé is the proud but fretful owner of an extraordinary little bed and breakfast (*"pousada"* in Portuguese) that's located – well, I probably shouldn't tell. That's because Zé only sort of, kinda, halfway wants visitors, if you know what I mean. This is true even though he makes his living from having guests.

Zé and his wife, Fátima, built their little *pousada* at the very end of the road, at the very end of the power lines, at the very end of your patience to ever get there. They built it 30 kilometers beyond the last gas station and fruit stand. They built it on the very lip of one of the most beautiful white-sand beaches I've ever seen, fronted by one of the cleanest and most beautiful rivers in the world: the *Rio Tapajós*. (I don't feel like I'm breaching confidence here in mentioning the *Tapajós*. Since the river is more than 1,000 miles long, it affords Zé plenty of room to hide.)

Though Ze's *pousada* has yet to be mentioned in any travel guide ever printed, he plays host to a strange and loyal bunch of patrons (including a pair of tired missionaries come looking for a rest). Some of Zé's clients travel thousands of miles – from Buenos Aires and Paris and Montreal – to reach this place, and they, like Zé, would just as soon it remain *escondida*.

"I need to move someplace farther out," he mused again. Since we were sitting in a place that is lost in time and space, it made no difference to us that 15 minutes had lapsed since he'd first expressed the notion.

"Why move?" I asked him. "It's paradise right here."

He nodded grimly. "You know that road you came down to get here?"

I smiled at the thought. The *"road"* he referred to is a raw gash of washboard and soft sands, bulldozed through 20 kilometers of scrub jungle. Who knows what kind of monster it becomes during the rainy season? It will take someone with a lot more courage and car than mine to find out.

"Sure," I said. "*The road.*"

"They're gonna pave it," he said flatly.

"*Sure* they will."

"No, really," he assured me. "It was promised in the last elections."

"And you *believe* that?"

"I do. 'Cause there's money in it for them. And when they do pave it, everything – both good *and* bad – is gonna come driving down that road and land right here on my veranda. And this paradise we're sitting in – right here – is gonna change."

"It'll bring you customers, Zé. Lots of customers."

"That'll be a blessing," he smiled. But his lips couldn't hold it long. "And *also* a curse." His eyes ran up the beach a ways, and back again, and found nothing there to break their lazy run but a pair of herons and a fisherman tending his nets.

"Maybe you'll come too?" he suggested. "Think about it, gringo. Think about *someplace farther out*."

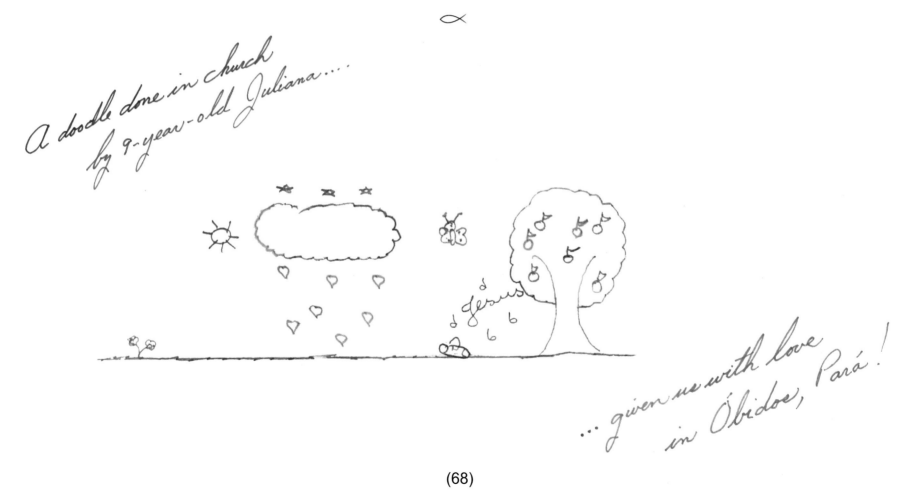

A doodle done in church by 9-year-old Juliana....

... given us with love in Óbidos, Pará!

Dream Seeds

The greatest man you'll ever be
Lies in the dream God dreams for thee.
There like a seed, within your soul,
Pray let it grow and make you whole!

Moonlight on the Amazon

Dear Clara,

I read your letter about the president and could feel the steam underneath it.

But I'm in no mood to talk politics this morning. Nor, I suspect, will I be in any better mood to talk politics this afternoon or tonight.

I am thinking instead about something tremendous that I saw out on the *Rio Ituquí* the night before last.

We were with some Brazilian friends – fishermen of great skill and integrity – who had offered to show us their neighborhood by night. Using a pair of shallow-bottomed canoes, they paddled us out into the vast marshes that border the river.

A curtain of heavy rain passed through, wetting our clothes and leaving a nervous amount of water sloshing around our feet. The distance between the top of the canoe's wooden gunwale and the surface of the marsh was only the length of my hand.

The moon was frail and setting early as we glided along together. It is a world that few are ever privileged to see – the Amazon by night.

Suddenly, as if some secret sign were given, our friends stopped paddling, turned their flashlights on and trained their beams along the shore. Like orange-hot coals stirred up from cool ash, the eyes of the gators gleamed back at us. Here a pair and there we saw them. Sometimes in groups of three or four, like clusters of pearls luminescing.

Our friends say they can tell the length of a gator by gauging the size of its eyes and the distance they're set apart. This is a life-and-death skill for the native hunters, who go in with simple spears, knowing that in among the babies and the adolescents are some fully armored monsters – six and seven meters long – that will take your leg off before you have time to recalculate your error.

I have seen a great many colors from God's fine palette, in His sunsets and peacock tails, His desert rocks and orchids. But never have I seen an orange like that which looked me over on the *Rio Ituqui*.

And so, all politics aside, I pose you this question, Clara:

Who put the fire in the alligator's eyes?

Seriously yours,

Share the Moonlight!

Please tell someone you love about

Moonlight on the Amazon

or better yet, send them a copy!

*With your help we hope to generate support
for world missions and ministries to the poor.*

To purchase copies, please visit

www.moonlightontheamazon.com

or www.amazon.com.

∝